Mahwish Fatima
Kiran Shahzadi
Shazia Shereen

Effect of Different Fertilizers on the Growth of Capsicum annuum

Mahwish Fatima
Kiran Shahzadi
Shazia Shereen

Effect of Different Fertilizers on the Growth of Capsicum annuum

Effect of Organic and Inorganic Fertilizers on the Growth of Sweet Pepper(Capsicum annuum L.)

LAP LAMBERT Academic Publishing

Imprint

Any brand names and product names mentioned in this book are subject to trademark, brand or patent protection and are trademarks or registered trademarks of their respective holders. The use of brand names, product names, common names, trade names, product descriptions etc. even without a particular marking in this work is in no way to be construed to mean that such names may be regarded as unrestricted in respect of trademark and brand protection legislation and could thus be used by anyone.

Cover image: www.ingimage.com

Publisher:
LAP LAMBERT Academic Publishing
is a trademark of
International Book Market Service Ltd., member of OmniScriptum Publishing Group
17 Meldrum Street, Beau Bassin 71504, Mauritius

ISBN: 978-3-659-12177-7

LIST OF CONTENTS

Chapters	TITLE	Page no

Chapters	TITLE	Page no
	List of Tables	2
	List of Plates	3
	Dedication	4
	Abstract	5
1	Introduction	6
2	Review of Literature	5
3	Materials and Methods	22
4	Results and discussion	29
5	References	61

LIST OF TABLES

Table	Description	Page no
4.1	Effect of growing media on %age germination of sweet pepper	30
4.2	Pre-Harvest soil analysis.	33
4.3	Effect of different soil compositions on shoot lengths of sweet pepper.	35
4.4	Effect of different soil compositions on number of leaves of sweet pepper.	39
4.5	Effect of different soil compositions on chlorophyll content of sweet pepper.	42
4.6	Effect of different fertilizer composition on shoot length of sweet pepper.	44
4.7	Effect of different fertilizer compositions on number of leaves of sweet pepper.	48
4.8	Effect of different fertilizer compositions on chlorophyll content of sweet pepper.	51
4.9	Post harvest analysis of plants.	53
4.10	Post harvest soil analysis.	57

LIST OF PLATES

Table	Description	Page no
4.1	Effect of growing media on rate of germination of sweet pepper .	31
4.2	Effect of different soil compositions on shoot lengths of sweet pepper.	36
4.3	Effect of different soil compositions on number of leaves of sweet pepper .	40
4.4	Effect of different Fertilizers on shoot length of sweet pepper.	45
4.5	Effect of different Fertilizers on number of leaves of sweet pepper.	49
4.6	Post harvest analysis of plants.	54

DEDICATIONS

I dedicated this effort to my respected Parents Syed Tauqir Abbas Kazmi and Mrs. Najaf Naqvi and to my aunt Narjis Naqvi and my sister Kiran Fatima.

ABSTRACT

The present study was conducted to observe the effects of different growing media on germination and different soil media on the growth of *Capsicum annuum* L. Seeds of sweet pepper were firstly sown in pots with bhal where the percentage of germination was observed. Then the seedlings were transplanted in the pots in Green House. There were five treatments T0 (100% bhal), T1 (50% bhal + 50% leaf manure), T2 (100% biofertilizer), T3 (50% bhal + 50% biofertilizer) and T4 (50% bhal + 50% organic manure). Vegetative growth of potted sweet pepper plants was monitored during the course of experiment by measuring various growth parameters viz. plant height and number of leaves. Both the parameters were less affected with the biofertilizer T2, T3 &T4 and the growth was maximum in T1. Then the seedlings were grown in the soil composition that showed the maximum results i.e. T1. Their soil analysis was done with various parameters i.e. pH, electrical conductivity, estimation of CO_3^{2-} and HCO_3^{1}. Plant height and number of leaves were less affected with greater concentration of bhal i.e. T0 and growth was maximum in T1. Then the plants with the T1 were fertilized with different composition of fertilizers i.e. Phostrogen (2.75, 3.75, 4.75) g/l, NPK (0.927, 1.85, 3.75) g/l and Hoagland (50, 100, 150) ml/l, whose soil analysis was done with various parameters i.e. pH, electrical conductivity, estimation of CO_3^{2-} and HCO_3^{1-}. Plant height and number of leaves were maximum in NPK (3.75 g/l). Plants of all the treatments were harvested at the prime of their vegetative growth in order to assess the soil media impact on them. It was found that plant fresh and dry weight was greater in NPK (3.75 g/l). Similarly chlorophyll a, b and total chlorophyll content showed maximum results in NPK (3.75 g/l). Finally post harvest soil analysis was carried out for the same parameters. It was observed that pH, electrical conductivity, CO_3^{2-} and HCO_3^{1-} values were maximum at NPK (3.75 g/l).

INTRODUCTION

Pepper, also called Garden Pepper, (*Capsicum*) belongs to the nightshade family, Solanaceae, notably *Capsicum annuum, C. frutescens,* and *C. boccatum,* extensively cultivated throughout tropical Asia and equatorial America for their edible, pungent fruits. Peppers, which have been found in prehistoric remains in Peru, were widely grown in Central and South America in pre-Columbian times. Pepper seeds were carried to Spain in 1493 and from there spread rapidly all over Europe. (www.britannica.com). Pepper, native to the tropics of Central and South America, has probably been cultivated for thousands of years. South America, Spain, England and the Caribbean all played roles in the introduction of the pepper to North America (http://aggie-horticulture.tamu.edu/archives/parsons). In his journal Columbus faithfully recorded his sighting of a new pungent, red-fruited plant that he called pepper, and he brought back specimens to Spain, marking the beginning of the history of capsicums for the people of the Old World. (www.answer.com/topic/chilli-pepper).

Bell pepper or sweet pepper is a cultivar group of the species *Capsicum annuum* (chili pepper).Cultivars of the plant produce fruits in different colors, including red, yellow and orange. The fruit is also frequently consumed in its unripe form, when the fruit is still green. Peppers are native to Mexico, Central America and northern South America. Today, Mexico remains one of the major pepper producers in the world. Green peppers are less sweet and slightly bitterer than red, yellow or orange peppers. The taste of ripe peppers can also vary with growing conditions and post-harvest storage treatment; the sweetest are fruit allowed to ripen fully on the plant in full sunshine, while fruit harvested green and after-ripened in storage are less sweet. (en.wikipedia.org/wiki/capsicum_annuum).

Capsicum annuum L. and *Capsicum frutescens L.,* plants are used in the manufacture of selected commercial products known for their pungency and color. *Capsicum annuum L.* is herbaceous, perennial but usually grown as annual that reaches a height of one meter and has glabrous or pubescent lanceolate leaves, white flowers, and fruit that vary in length, color, and pungency depending upon the cultivar (www.calantilles.com). The plant reaches 0.5–1.5 m (20–60 inch). Single white flowers bear the fruit which is green when unripe, changing principally to red, although some varieties may ripen to other colors including brown and purple. Sturdy plants grow 30-36" tall and produces 3-4" fruits

6

similar in shape to a jalapeno, but black in color ripening to red (en.wikipedia.org/wiki/capsicum). Capsicums are mainly self-pollinating and do not need any insect vector for pollen transfer while according to some scientists insect pollination has been considered to improve fruit quality and thus increase the incomes of farmers (Burt, 2005). There is little evidence to suggest that capsicums are fully pollinated in the absence of insect pollinators, thus it can be assumed that insects may play a significant yet unconfirmed role in capsicum production. It has been demonstrated that the weight and size of fruit and the number of seeds they contain are directly proportional to pollen load in various species. Flowers always open by day break. The length of time flowers remain open is dependent on temperature and humidity. If temperature is low (10-12.8°C) and the humidity is over 75%, the flowers may remain open until midday. Sweet pepper is a warm and dry season crop. It germinates well at temperature of 20°C to 30°C and grows best at around 25°. It can be grown throughout the year in mid and high elevations. Sweet pepper grows well in any type of soil with pH of 5.5 to 6.5. Production is best, however, in deep loam soil with good fertility, easy irrigation, adequate drainage and plenty of sunshine. It is best to rotate the crop with rice, legume, sugarcane and corn (www.da.gov.ph/tips/sweetpepper.html).

Sweet pepper is a nutritious vegetable that is gaining in popularity throughout the world. According to Irfan & Khalid (2007) Sweet pepper belongs to the family Solanaceae which is an important group of vegetables cultivated extensively in Pakistan & also widely cultivated in almost every country of the world. Sweet pepper is the summer crop & its total area under cultivation in Pakistan is about 91800 hectare, with total annual production of 115,000 tones. Research is now being conducted to increase this vegetable's adaptability and tolerance to heat in the tropics. (www.avrdc.org/LC/pepper/swtprod/03int.html).

Cayenne or Capsicum takes its name from the Greek, 'to bite,' a reference to the hot pungent properties of the fruits and seeds. The fruits contain 0.1 - 1.5% capsaicin, a substance which stimulates the circulation and alters temperature regulation. The seed contains capsicidins, which are thought to have antibiotic properties. The fruit can be used raw or cooked. It is normally used as a flavoring. The fruit can also be dried and ground into a powder for use as a flavoring. The seeds are also dried, ground and used as a pepper and the leaves can be cooked as a potherb. The dried fruit has no narcotic effect but as a

7

powerful local stimulant, it is effective in dilating blood vessels and relieving chronic congestion. Due to its high vitamin C content, it is said to be good at warding off disease. (www.wisegeek.com). *Capsicum annuum L.* is used in sweet bell peppers, paprika, pimento, and other red pepper products. The extracts of Capsicum species have been reported to have antioxidant properties. As a medicinal plant, the Capsicum species has been used as a carminative, digestive irritant, stomachic, stimulant and tonic. The plants have also been used as folk remedies for dropsy, colic, diarrhea, asthma, arthritis, muscle cramps, and toothache (www.calantilles.com/capsicum_pepper). John Lindley wrote in his 'Flora Medica' about *Capsicum annuum* 'it is employed in medicine, in combination with Cinchona in intermittent & lethargic affections and also in atonic gout, dyspepsia accompanied by flatulence, tympanitis, paralysis etc. For people worried about colon cancer, the fiber found in peppers can help to reduce the amount of contact that colon cells have with cancer-causing toxins found in certain foods or produced by certain gut bacteria. In addition, consumption of vitamin C, beta-carotene, and folic acid, all found in bell peppers, is associated with a significantly reduced risk of colon cancer. The red varieties of bell peppers also supply the phytonutrients lutein and zeaxanthin, which have been found to protect against macular degeneration, the main cause of blindness in the elderly (www.whfoods.com/bellpepper).

Pepper plants should be fertilized with a combination of organic (animal manure) and/or chemical fertilizers to produce high yields. Soil conditions and cropping patterns should be considered when selecting fertilizer amounts and conducting a soil test is strongly recommended (www.avrdc.org/LC/pepper/swtprod/09fert.html). Peppers require a fertilizer to help give them a push during their long growing season. Fertilizer with an even mix of nitrogen, phosphorus and potassium (NPK) should be added to indoor pepper plants after they have germinated (pushed through the soil). By using a fertilizer NPK, nitrogen helps boost a plant's foliage growth above ground. Plants with lacking nitrogen appear stunted. Phosphorus is required by a plant for the conversion of light energy to chemical energy during photosynthesis and also for cell communication and reproduction. It helps roots, fruit and flower development. Plants with lacking Phosphorus show signs of stunted growth, taking longer to mature. Potassium regulates the water transfer through the plant, reducing water loss from the leaves, making them more resistant to cold and dry weather. Unfortunately deficiencies only become visible when they are severe (www.thechileman.org/guide-fertilizer.php)

Phostrogen Plant Food is a soluble fertilizer for feeding all garden and pot plants. Phostrogen's high potash formulation, with trace elements for root and foliar feeding, is recommended where flowering, fruiting, hardiness and disease resistance are important Phostrogen Plant Food contains optimum levels of nitrogen, phosphate and potash. Phostrogen is easy to apply by watering can or Phostrogen feeder (www.debco.com.au/products/fertilisers/phost.php).

The Hoagland solution is a hydroponic nutrient solution that was developed by Hoagland and Snyder in 1933 and is one of the most popular solution compositions for growing plants (in the scientific world at least). The Hoagland solution provides every nutrient necessary for plant growth and is appropriate for the growth of a large variety of plant species. The Hoaglands solution has a lot of N and K so it is very well suited for the development of large plants like Bell Pepper. (http://en.wikipedia.org/wiki/Hoagland_solution).

Keeping in view the importance of Sweet pepper in our life as a spice & its other usage, the research has been directed to evaluate the best growing media, soil composition & best fertilizers for this most important crop. The aim of this study is to evaluate the effect of different soil compositions and fertilizers on the growth of the Sweet pepper. These compositions were Biofert 100%, Bhal 100%, Bhal + Leaf manure (50%+50%), Bhal + Organic manure (50%+50%), Bhal + Biofertilizer (50%+50%). The different fertilizers we used in this experiment were Soluble NPK (0.927g, 1.87 and 3.75g make volume up to liter with distilled water), Phostrogen (2.75g, 3.75g and 4.75g make the volume up to liter) and Hoagland solution (50ml stock solution, 100ml stock solution and 150ml stock solution make the volume up to liter). Hence the objective of this experiment were to find the optimum ratio of growth medium giving the best growth & fruit yield of Sweet pepper & to find the effectiveness of different soil compositions & effectiveness of different fertilizers on the morphological characters of the plants.

REVIEW OF LITERATURE

Solanaceae is commonly called as 'Potato family' or nightshade family. It includes about 100 genera and about 2,500 species. Members having a cosmopolitan distribution and are mostly herbs, undershrubs and shrubs; some are rarely trees & climbers. It is an important source of food, spice and medicine. It is also used for ornamental purposes. (www.tutorvista.com). Important genera of this family with economical importance as food and medicine plants are the potato (*Solanum tuberosum*); brinjal or egg plant (*Solanum melongena*); tomato (*Lycopersicon esculentum*); sweet pepper or capsicum (*Capsicum annuum*); hot pepper (*C. frutescens*); tobacco (*Nicotiana tabacum*); deadly nightshade, the source of belladonna (*Atropa belladonna*); the poisonous jimsonweed *(Datura stromonium)* and night shades (*S. nigrum, S. dulcamara, and others*); and many garden ornamentals, such as the genera *Brugmansia, Cestrum, Nicandra, Nicotiana, Nierembergia, Petunia, Salpiglossis, Schizanthus, Solandra, Solanum , Datura, Browallia, and Burnfelsia.* . (www.britannica.com).

Sweet pepper refers to a wide variety of tropical pepper plants, belong to Capsicum genus. Common names for these plants are cayenne, cayenne pepper, chili pepper, paprika, peppers, pimiento, red pepper, sweet pepper, aji dulce, Hungarian pepper, and Mexican pepper. (www.wisegeek.com/capsicum). Plant produces fruits in different colors, including red, yellow, and orange. The fruit is also frequently consumed in its unripe form, when the fruit is still green. Peppers are native to Mexico, Central America and northern South America. Pepper seeds were later carried to Spain in 1493 and from there spread to other European, African and Asian countries. Today, Mexico remains one of the major pepper producers in the world. (www.myfolia.com/capsicum).

Scientific name: *Capsicum annuum*
Family: Solanaceae
English name: Pepper, Paprika or sweet pepper.
Hindi name: Simla Mirch.
(en.wikipedia.org/wiki/capsicum).

Melo, (1997) reported that Sweet pepper (*Capsicum annuum L.*) is one of the most consumed vegetables in Brazil, and is better adjusted and most frequently produced under greenhouse conditions. Cultivation of sweet pepper in greenhouses demands intensive use of agri-inputs.

Table # 1: Constituents of sweet pepper as per 100g of edible portion.

Constituents in sweet pepper	Nutritional values per 100g
Energy	20 kcal
Carbohydrates	4.64g
Sugars	2.40g
Dietary fiber	1.7g
Fat	0.17%
Protein	0.86%
Thiamin (Vit. B1) 0.057mg	4%
Riboflavin (Vit. B2) 0.028mg	2%
Niacin (Vit. B3)0.480 mg	3%
Pantothenic acid (B5) 0.099 mg	2%
Vitamin B6 0.224 mg	17%
Folate (Vit. B9) 10 µg	3%
Vitamin C 80.4 mg	134%
Calcium 10mg	1%
Iron 0.34mg	3%
Magnesium 10mg	3%
Phosphorous 20mg	3%
Potassium 175mg	4%
Zinc 0.13mg	1%

(www.nal.usda.gov/fnic/foodcomp/search)

Boukhenfouf and Ahmed (2011) studied that because of their mineral content, soils are naturally radioactive and one of the sources of radioactivity other than those of natural origin is mainly due to the extensive use of fertilizers. The main aim was to evaluate the fluxes of natural radionuclide in local production of phosphate fertilizers to determine the content of radioactivity in several commercial fertilizers produced in Algeria and to estimate their radiological impact in a cultivated soil even for the long-term exposure due to their application. For these purposes, virgin and fertilized soils were collected from outlying Setif region in Algeria and from phosphate fertilizers used in this area.

Russo, (2010) reported that excessive rain fall might leach nutrients from the soil or cause producers to not supply irrigation to pepper (*Capsicum sp.*). Fertilizer at 150 or 300 lb/acre of triple 17 NPK, the lower rate is the recommended rate, was supplied to either bell, cv. Jupiter, or non-pungent jalapeno, cv. Pace 105, peppers, both *C. annuum* L. Irrigation was either not applied, or the same amount of irrigation was applied either once or twice weekly at times in the day when air temperatures would be different, 10AM or 2PM. Fertilizer rate and time of day when irrigation was applied did not affect yield. Irrigation produced marketable yields that were 2 to 3 times greater than no irrigation. Irrigation twice a week produced marketable yields for bell pepper that were 2.2 times greater than irrigation once a week. For non-pungent jalapeno peppers more frequent irrigation produced yields that were 1.4 times greater. Additional fertilizer is not, but irrigation is, necessary to maintain yields even when rainfall levels are above normal.

Pascual *et al.*, (2010) used organic wastes such as sewage sludge successfully to increase crop productivity of horticultural soils. Nevertheless, considerations of the impact of sludges on vegetable and fruit quality have received little attention. Therefore, the objective of the present work was to investigate the impact of two sanitized sewage sludges, autothermal thermophilic aerobic digestion and compost sludge, on the growth, yield, and fruit quality of pepper plants (*Capsicum annuum L.* cv. Piquillo) grown in the greenhouse. Two doses of autothermal thermophilic aerobic digestion (15 and 30% v/v) and three of composted sludge (15, 30, and 45%) were applied to a peat-based potting mix. Unamended substrate was included as control. Autothermal thermophilic aerobic digestion and composted sludge increased leaf, shoot, and root dry matter, as well as fruit yield, mainly due to a higher number of fruits per plant. There was no effect of sludge on fruit size (dry matter per fruit and diameter). The concentrations of Zn and Cu in fruit increased with the addition of sewage sludges. Nevertheless, the levels of these elements remained below toxic thresholds. Pepper fruits from sludge-amended plants maintained low concentrations of capsaicin and dihydrocapsaicin, thus indicating low pungency level, in accordance with the regulations

12

prescribed by the Control Board of "Lodosa Piquillo peppers" Origin Denomination. The application of sludges did not modify the concentration of vitamin C in fruit, whereas the highest doses of composted sludge tended to increase the content of reduced and oxidized glutathione, without change in the reduced and oxidized glutathione ratio. The application of sanitized sludges to pepper plants can improve pepper yield without loss of food nutritional quality, in terms of fruit size and vitamin C, glutathione, and capsaicinoid contents.

Amor et al., (2010) reported the effect of different air temperatures (10, 20 and 30 °C) on the response of sweet pepper plants (*Capsicum annuum L.* cv. Herminio) to foliar urea applications after growing plants for 20 day with and without nitrogen (N) applied to the growing substrate. Leaf CO_2 assimilation, chlorophyll fluorescence, root respiration, lipid peroxidation and antioxidative enzymes were analysed. Spraying plants with urea increased leaf CO_2 assimilation of N-deficient plants when applied at 20 or 30 °C, compared with non-sprayed plants. When plants were sprayed with urea at 10 °C chlorophyll fluorescence of leaves was similar to that of plants that were supplied with full N in the nutrient solution. Root respiration was not affected by urea sprays at the same time as leaf NO_3^- concentration was increased by urea but only when it was sprayed at 10 or 20 °C. Lipid peroxidation and ascorbate peroxidase in N-deficient plants were reduced significantly by urea sprays, especially when plants were sprayed at 20 °C with N-limitation in the growing substrate. This study shows that N-limitation in the growing substrate induces a temperature-dependant increase in the activities of antioxidant enzymes in leaves of pepper and applications of foliar urea can be optimised, when applied at the suitable temperature, to partly replace the N supplied to the roots of sweet pepper.

Kotur et al., (2010) conducted a field experiment in a sandy loam soil involving 'Arka Lohit', a high-yielding variety and an F1 hybrid chilli (*Capsicum annuum L.*) during 2002–05 using 15N-enriched (1% N abundance) urea as the tracer to evaluate direct and residual of applied nitrogen in chilli–radish (*Raphanus sativus* L.) system. Response in terms of dry matter production was observed in both the varieties when 100% N dose (120 and 200 kg/ha, respectively) was applied in 3 splits, i.e. basal + 2 top-dresses. The highest yield 'Arka Lohit' was obtained when the same dose was applied in 2 equal splits as basal + top-dress due to its slower growth habit. F1 hybrid chilli hybrid out yielded other treatments when 100% N dose was applied in 3 splits as basal + 2 top-dresses due to its quick yielding habit. Either deferring the time of application by 10 days or reducing the N dose by 20–40% and increasing the splits from 2 to 3 were not beneficial.

Dileep and Sasikala, (2009) carried out a study to find out the effect of different sources of organic manures along with various levels of inorganic fertilizers on growth, fruit traits, yield and quality improvement of chilli cv. K1. Chilli is one of the commercial high value crops in their country. Chilli crop requires a balanced fertilizer management without which growth and development of the crop will be impaired leading to considerable reduction not only in yield but also in the market appeal of the produce namely the color and quality of the dry chilli. From the study, it was found that growth, yield and quality attributes of chilli were significantly influenced by different treatment combinations. Literature may also be concluded that application of 75% recommended dose of fertilizers along with humic acid 30 kg/ha can help to increase the growth, yield and quality of chilli cv. K-1.

Gopinath et al., (2009) studied that a conversion period of at least two years is required for annual crops before produce may be certified as organically grown. There is a need for better understanding of the various management options for implementing from conventional to organic production. The purpose of this study was to evaluate the effects of three organic amendments on growth and yield of bell pepper (*Capsicum annuum L.*), the benefit: cost ratio, soil fertility and enzymatic activities during conversion to organic production. For that purpose six treatments were established: composted farmyard manure (FYMC, T_1); vermicompost (VC, T_2); poultry manure (PM, T_3) along with biofertilizer (BF) [*Azotobacter* + phosphorus solubilizing bacteria (*Pseudomonas striata*)] and mix of three amendments (FYMC + PM + VC + BF, T_4); integrated crop management (FYMC + NPK, T_5) and unamended control (T_6). The bell pepper yield under organic management was markedly lower (33–53% and 18–40% less in first and second year of conversion, respectively) than with the integrated crop management (FYMC 10 Mg ha^{-1} + NPK – 100:22:41.5 kg ha^{-1}) treatment (T_5). Combined application of three organic amendments (FYMC 10 Mg ha^{-1} + PM and VC each 1.5 Mg ha^{-1} + BF, T_4) and T_1 produced similar but significantly higher bell pepper yield (27.9 and 26.1 Mg ha^{-1}, respectively) compared with other organic amendment treatments. Both T_4 and T_1 greatly lowered soil bulk density (1.15–1.17 Mg m^{-3}), and enhanced soil pH (7.1) and oxidizable organic carbon (1.2–1.3%) compared with T_5 and unamended control (T_6) after a two-year transition period. However, the N, P and K levels were highest in the plots under integrated management. T_1 plots showed

14

higher dehydrogenase activity values. However, acid phosphatase and β-glucosidase activities were higher in T_6 plots whereas urease activity was greater in T_5 plots compared with other treatments. Among the treatments involving organic amendments alone, T_1 gave a higher gross margin (US $ 8237.5 ha^{-1}) than other treatments. Hence, T_1 was found more suitable for enhancing bell pepper growth and yield, through improved soil properties, during conversion to organic production.

Serrano et al., (2009) applied a mix of nitrophenolates to pepper plants in the irrigation system along the growth cycle. Fruits were labelled at fruit set to study the evolution of fruit growth and ripening based on fruit size and colour. In addition, at 3-day intervals, samples were taken in which the development of fruit weight, colour, nutritive (sugars and organic acids) and bioactive compounds (total phenolics, carotenoids, and ascorbic acid) was evaluated. Pepper fruit growth followed a simple sigmoid curve reaching its maximum size at 49 days after fruit set, although nitrophenolate treatments led to significant increases in fruit weight due to higher length, diameter, and pericarp thickness, without affecting the normal ripening process, since colour and carotenoid evolution was similar for both control and treated fruits. Glucose, fructose, ascorbic acid, citric acid, total antioxidant activity and total phenolics increased during pepper development, and their levels were significantly enhanced by nitrophenolate applications. Thus, this treatment induced beneficial effects in terms of the improvement of fruit quality, and especially its nutritive and antioxidant constituents. Finally, it is advisable to consume peppers at the full red stage in order to achieve the maximum health-beneficial effects by consumers.

El-Al F. S. A, (2009) conducted two field experiments during the two successive seasons of 2006 and 2007 to investigate the effect of each of some organic acids (gibberellic acid, salsillic acid and citric acid) and the urea application method on the growth, fruits, yield and its physical and chemical consistent of sweet pepper. The obtained results indicated that the soil dressing of urea fertilizer resulted in vigor plant growth of sweet pepper as expressed by plant length, number of leaves and shoots, fresh and dry weight of different plants organs and gave the heavier total fruits yield and the better physical quality and chemical quality. The soil dressing is more benefit for plant compared with the foliar spraying. The foliar spraying of sweet pepper by gibberellic acid gave the best plant growth and the heaviest tonnage of total, early fruits yield and its physical and chemical consistent of fruits.

15

Berova and Karanatsidis, (2009) conducted an experiment in which pepper plants were grown in a phytostatic chamber under controller conditions. Seedlings were fertilized with a bio-fertilizer, produced by the Californian earthworm *Lumbricus rubellus* at concentrations of 50.0 and 100.0 ml/per plant respectively. It was discovered that biofertilizer speeds up plant growth. It influences the growth rate of the roots and stems and affects the formation of the leaf area. There is a positive effect of the bio-fertilizer upon the functional activity of the photosynthetic apparatus increased content of photosynthetic pigments, improved leaf gas exchange.

Marin *et al.,* (2008) investigated that integrated, organic, and soil-less production systems are the principal production practices that have emerged to encourage more sustainable agricultural practices and safer edible plants, reducing inputs of plaguicides, pesticides, and fertilizers. Sweet peppers grown commercially under integrated, organic, and soil-less production systems were compared to study the influence of these sustainable production systems on the microbial quality and bioactive constituents (vitamin C, individual and total carotenoids, hydroxycinnamic acids, and flavonoids). The antioxidant composition of peppers was analyzed at green and red maturity stages and at three harvest times (initial, middle, and late season). Irrigation water, manure, and soil were shown to be potential transmission sources of pathogens to the produce. Coliform counts of soil-less pepper were up to 2.9 log units lower than those of organic and integrated peppers. Soil-less green and red peppers showed maximum vitamin C contents of 52 and 80 mg 100 g^{-1} fresh weight respectively, similar to those grown in the organic production system. Moreover, the highest content of total carotenoids was found in the soil-less red peppers, which reached a maximum of 148 mg 100 g^{-1} fw, while slightly lower contents were found in integrated and organic red peppers. Hydroxycinnamic acids and flavonoids represented 15 and 85% of the total phenolic content, respectively. Total phenolic content, which ranged from 1.2 to 4.1 mg 100 g^{-1} fw, was significantly affected by the harvest time but not by the production system assayed. Soil-less peppers showed similar or even higher concentrations of bioactive compounds than peppers grown under organic and integrated practices. Therefore, in the commercial conditions studied, soil-less culture was a more suitable alternative than organic or integrated practices, because it improved the microbial safety of sweet peppers without detrimental effects on the bioactive compound content.

identified promising lines of chili rootstock for production of grafted sweet pepper during the hot-wet and hot-dry seasons in the lowland tropics.

Alabi, (2005) monitored the effect of five phosphorus levels (0, 25, 50, 75, 100 and 125 kg/ha) and five poultry dropping (0, 100, 200, 300, 400 and 500 kg/ha) levels on the growth, growth yield, yield components, nutrients concentration and food values of pepper (*Capsicum annuum* L) were observed from 2002 to 2003 raining seasons. Phosphorus levels significantly increased pepper plant height, number of leaves per plant, number of branches per plant and leaf area up to 125 kg (p/ha) level. The phosphorus application also significantly increased early flowering, maturity and yield (ton/ha) of the treated plants. Application of organic waste, poultry dropping increased the growth yield and yield components of pepper significantly more than the fertilizer phosphorus. Since poultry dropping is cheap to obtain and enhances the pepper plant performance, it is recommended for use. *C. annuum* was found to contain both the major and minor nutrient elements which can supply the body with necessary ingredients for growth.

Russo, (2005) reported the effectiveness of using potting media and fertilizers that are alternatives to conventional materials to produce vegetable transplants needs clarification. Bell pepper, onion and watermelon seed were sown in Container Mix, Lawn and Garden Soil, and Potting Soil, which can be used for organic production in greenhouse transplant production. The alternative media were amended with a 1× rate of Sea Tea liquid fertilizer. Comparisons were made to a system using a conventional potting medium, Reddi-Earth, fertilized with a half-strength (0.5×) rate of a soluble synthetic fertilizer Peters. Watermelon, bell pepper and onion seedlings were lifted at 3, 6, and 8 weeks, respectively, and heights and dry weights determined. Watermelons were sufficiently vigorous for transplanting regardless of which medium and fertilizer was used. Bell pepper and onion at the scheduled lifting were sufficiently vigorous only if produced with conventional materials. seedling development.

Marcussi *et al.,* (2004) reported that information on nutrient demand during each growth stage is essential for efficient application of nutrients. A pot experiment was carried out with a Typic Hapludox under greenhouse conditions in Botucatu, SP, Brazil, aiming was to determine nutrient uptake and partition of sweet pepper plants, cultivar Elisa in randomized block design with four replications. The fertigation was simulated through 2-L PET bottles (neck down with a tube and a flow regulator at the end, simulating a drip irrigation system). Four plants per replication were collected at eight growth stages (0, 20, 40, 60, 80, 100, 120

19

and 140 days after the seedling transplant - DAT). The period of largest extraction of nutrients for the plant occurred from 120 to 140 days after the seedling transplant, which coincides with the highest accumulation of dry phytomass. The highest Mg and Ca accumulation occurred in the leaves, while N, K, S and P were mostly accumulated in the fruits. Only 8 to 13% of the total amount of the accumulated macronutrients at 140 days after the seedling transplant was absorbed up to the 60th days after the seedling transplant. Between the 61st and 100th days after the seedling transplant, K was the most absorbed macronutrient (60% of the macronutrients accumulated during the whole cycle). P, Ca and S were the most absorbed nutrients at the end of the cycle. Considering rates (g per plant), the most absorbed macronutrients were: N (6.6) > K (6.4) > Ca (2.6) > Mg (1.3) > S (1.1) > P (0.7).

Gowda *et al.,* (2002) conducted a field experiment to study the effect of nitrogen and phosphorus fixing biofertilizer at various levels of nitrogen and phosphorus on growth, yield and quality parameters of red chillies cv. Byadagi Dabba at University of Agricultural Sciences, Gandhi Krishi Vignana Kendra, Bangalore (India). The maximum plant height, number of branches per plant, leaf area and dry matter production per plant were recorded in plant supplied with 75 per cent nitrogen, phosphorus plus 100 per cent potassium in addition to the inoculation of Azotobacter, Azospirillum, Phosphate solubilizing bacteria and Vesicular arbuscular mycorrhizas. Besides, the same treatment recorded more number of fruits per plant, fruit length, and fruit girth, number of seeds per fruit, dry weight of hundred fruits and higher yield of dry chillies. The maximum total suspended solids, ascorbic acid, oleoresin and capsaicin content in dry chillies were also maximum at this level. Application of bio-fertilizers along with reduced levels of chemical fertilizers has beneficial effects compared to application of recommended NPK or bio-fertilizers alone.

Sui *et al,* (2002) studied the effect of fertilization on the qualities of sweet pepper (*Capsicum annuum.* L) cultured in greenhouse, the experiment was carried out by drop irrigation. The results showed that the most sensitive organ of sweet pepper for nitrate was lateral branch, in which the nitrate content was highest. The nitrate content in leaf was lower than that in lateral branch, and in fruit, the nitrate content was the lowest. High fertilization of N, P and K had no negative influence on content of nitrate and total soluble sugars in fruit. The content of soluble sugars in fruit were about 50%-55% under all treatment and higher than that in leaf and lateral branch. Decreasing fertilization would cause increasing of the inosital and sucrose

20

contents. Glucose and fructose were the main soluble sugars in fruit. The metabolism of carbohydrates was weakened by high fertilization, and it expressed as restraining the accumulation of starch content in leaf. The content of starch was the highest in fruit. The lipid content was exceeding 3-4 times in the leaf than in the lateral branch and fruit, and perhaps, it was due to the higher content of chlorophyll in the leaf.

Salama and Zake, (2000) conducted field experiments during two successive autumn seasons of 1995/96 and 1996/97 at EL-Bostan, EL-Behera Governorate, Egypt, to study the effect of organic manures on the yield and quality of sweet pepper (*Capsicum annuum* cv. Gedeon F_1 hybrid) under unheated plastic houses. Plant growth parameters such as plant height, number of leaves and dry matter content, early yield and total yield were increased by using pigeon manure at a rate of 45 kg/540 m^2 during irrigation alone or mixed with chicken manure at a rate of 22.5 kg/540 m^2 during irrigation. The increase in total yield due to such treatments was ~23% over the control treatments. It could be recommended to use pigeon manure at a rate of 45 kg/540 m^2 during irrigation or pigeon manure at 22.5 kg + chicken manure at 22.5 kg/540 m^2 during irrigation to produce high yield and quality of sweet pepper under unheated plastic houses. There are no significant between all treatments on chlorophyll content, carotenoids, ascorbic acid and total soluble solids.

MATERIALS & METHODS

3.1 Materials:

Healthier seeds of sweet pepper were procured from the Punjab Seed Certification Department, and fertilizers from Gulberg market, Lahore. Seeds were screened for uniformity of size. All the shriveled, infected and empty looking seeds were discarded.

3.2 Methodology:

To perform the experiment following characters and parameters were observed:

3.2.1 Soil media:

Three different types of soil media were used which are:

1. Bhal
2. Leaf manure
3. Biofertilizer
4. Organic manure

3.2.2 Preparation of different soil treatments:

Five levels of soil combinations i.e. T0, T1, T2, T3 and T4 were prepared. T0 (100% bhal) was prepared by taking whole one pot of 7 inches diameter of bhal. Similarly T1 (50% bhal+50% leaf manure) was prepared by taking 50:50 ratio of bhal and leaf manure. T2 (100% biofertilizer) was prepared by taking whole one pot of 7 inches diameter of biofertilizer.T3 (50% bhal+50% biofertilizer) was prepared by taking 50:50 ratio of bhal and biofertilizer. T4 (50% bhal+50% organic manure) was prepared by taking 50:50 ratio of bhal and organic manure respectively.

Biochemical analysis:

3.3.1 Determination of soil pH before and after harvesting

3.3.2 Determination of soil electrical conductivity and soil salinity before and after harvesting

3.3.3 Determination of CO_3^{-2} & HCO_3^{-1} before and after harvesting

Vegetative analysis:

3.3.4 Percentage germination and rate of germination

3.3.5 Shoot length (cm)

3.3.6 Number of leaves

3.3.7 Shoot and root fresh weight (g)

3.3.8 Shoot and root dry weight (g)

3.3 Soil Analysis:

Soil combinations used in this study were analyzed for biochemical characteristics before transplant and after transplant at the end of experiment from soil saturation extract which was prepared by the following procedure:

a. Preparation of Soil Saturation Paste:

1. Firstly, crushed soil samples were thoroughly sieved.
2. Then 200g of soil was taken from it in a 1000ml plastic beaker and then distilled water was gradually added to the beaker. The soil was then thoroughly mixed by using spatula so that the soil saturation paste shows following characteristics:

- There was no free standing on the soil surface.
- It had a glistening appearance and if a cut was given to the paste and beaker was slightly tilted, the cut is filled up.
- If a little of the paste was taken on the blade of spatula and held edge wise. It falls down.

b. Achievement of Soil Saturation Extract:

23

a. Bucchner funnel and suction flask of 500 ml were thoroughly washed with distilled water and then the bucchner was fitted on the suction flask so that it becomes air tight.

b. A whatman filter paper of 11cm diameter was placed inside the bucchner funnel.

c. And the paste was poured over it.

d. The suction flask was connected with the suction pump and the soil saturation extract was collected drop wise in the suction flask.

e. The suction pump was then switched off when crack started appearing on the soil surface in the bucchner funnel.

f. The extract from the suction flask was then transferred to the plastic bottle.

3.3.1 Determination of pH of soil:

50 ml of soil saturation extract was taken in the beaker. Put the electrode of pH meter which is already switched on one hour before working, in the beaker containing soil saturation extract. Then the pH of different soil combinations was measured with the same procedure. The procedure was explained by Mc Keague (1978) and Mc Lean (1982).

3.3.2 Determination of Electrical Conductivity:

According to Richards (1954), the following procedure is described to calculate the electrical conductivity:

Saturation extract from different treatments was prepared by the above mentioned methods. Electrical conductivity was measured with EC meter.

3.3.3 Determination of CO_3^{-2} & HCO_3^{-1}:

After the determination of pH, electrical conductivity and salinity by soil saturation extract, the presence of carbonates and bicarbonates in each soil composition were determined by the procedure explained by Richard in 1954 using the formula:

meq/l of CO_3^{-2}=2A*x Normality of acid / volume of sample x 1000

meq/l of HCO_3^{-1}= (B*-2A) x Normality of acid / volume of sample x 1000

*A: Burette reading for CO_3^{-2}

*B: Burette reading for HCO_3^{-1}

Experimental Setup:

To perform the experiment, two setups were prepared to observe mentioned morphological characters of sweet pepper. These steps are:

3.3.4 Setup:

This method was designed to study the Percentage germination and rate of germination of Sweet pepper in growing media. Ten plastic pots of fourteen inches diameter were taken and were thoroughly cleaned. The holes at the bottom of the pots were partially closed with pebbles in order to prevent excessive drainage and loss of soil water. The pots were filled with bhal, thirty seeds were sown in each pot and then the seeds were covered with peat. After some days germination of seeds was noted. The Percentage germination and rate of germination of seeds were then calculated.

Schedule of various events during this method is as follows:

Date of sowing   12th January, 2011.

Date of first seed germination   21st January, 2011.

Complete germination   11th March, 2011.

Transplantation of seedlings   21st March, 2011.

3.3.4 Pot setup:

Twenty five plastic pots of seven inches diameter were taken and were thoroughly cleaned. The holes at the bottom of the pots were partially closed with pebbles in order to prevent excessive drainage and loss of soil water. Five treatments were made. The pots were labeled according to their respective treatments i.e. T0 (100% bhal) or T1 (50% bhal+50% leaf manure), T2 (100% biofertilizer), T3 (50% bhal+50% biofertilizer) or T4 (50% bhal+50% organic manure). Pots were filled with respective soil samples. Selected seedlings were transplanted to the pots. Pots were watered regularly throughout the season

25

until their maturity. Insects were controlled by spraying insecticides and weeds were removed manually because of their small number.

Second pot setup

Twenty plastic pots of fourteen inches diameter were taken and were thoroughly cleaned. The holes at the bottom of the pots were partially closed with pebbles in order to prevent excessive drainage and loss of soil water. Two treatments were made. The pots were labeled according to their respective treatments i.e. T0 (100% bhal) and T1 (50% bhal+50% leaf manure). Pots were filled with respective soil samples. Selected seedlings were transplanted to the pots. Pots were watered regularly throughout the season until their maturity. Insects were controlled by spraying insecticides and weeds were removed manually because of their small number.

Third pot setup

Mature plants from the previous setup with T1 composition were taken. Nine treatments of three fertilizers were made. The pots were labeled according to their respective treatments i.e. NPK (0.927g, 1.85g, 3.75g NPK), Phosrtogen (2.75g, 3.75g, 4.75g Phostrogen), and Hoagland solution (50ml, 100ml and 150ml of stock solution of hoagland). Fertilizers were given to the plants with the interval of three days. Insects were controlled by spraying insecticides and weeds were removed manually because of their small number.

3.3.5 Growth Assessment:

Two main harvests were taken. Harvest 1 was taken when plant had completed their vegetative growth after one month of complete germination. Harvest 2 was taken after two months of complete germination.

3.4.1 Harvest 1:

At the prime of vegetative growth, chlorophyll determination was carried out.

Chlorophyll determination:

Chlorophyll determination was carried out using fresh leaves of main stem of various treatments. The absorbance was taken using UV- spectrometer.

The chlorophyll a, b and total chlorophyll were determined according to the method of Arnon (1949). Leaf samples were first of all roughly crushed in pestle and mortar. The crushed samples were ground thoroughly with small amount of 80% acetone for 3-5 minutes. The acetone was filtered and filtrate was collected in a beaker. Whole process was repeated with two to three washings of mortar using 2ml of 80% acetone. The extract was used to get wavelength on a spectrophotometer. One cuvette was filled with the extract solution and the other one was filled with 80% acetone for reference purpose. The absorbance at 663nm and 645nm was noted.

The chlorophyll a, b and total were calculated by following formula:

Chl a: [12.7(OD663)-2.69(OD645)] V/1000 X W

Chl b: [22.9(OD645)-4.68(OD663)] V/1000 X W

V= Volume of the extract (ml)

W= Weight of fresh leaves (g)

3.4.2 Harvest 2:

At the approach of crop maturity, second harvest was taken. The data is shown in the table. Following parameters were studied:

- Numbers of flowers per plant
- Shoot length (cm)
- Root length (cm)
- Fresh shoot weight (g)
- Fresh root weight (g)
- Total fresh weight of the plant (g)
- Dry shoot weight (g)
- Dry root weight (g)
- Total dry weight of the plant (g)

Statistical analysis of data:

A completely randomized design with 5 replicates was used for the experiment. The data for each parameter were subjected to analysis of variance (ANOVA) using the COSTAT V.63: statistical software (Cohort software, Berkely, California). The mean values were compared with the least significant difference test following Duncan's new multiple range test at 5% level.

RESULTS

Effect of growing media on rate of germination:

Rate of germination was also observed from first seedling till the last seed germinated. It was observed that there is gradual increase in rate of germination in bhal. Rate of germination was high on 14th day of germination in which 10 seeds germinated in pot 1 and 8 pots in pot 2 as mentioned in the table 4.1 and plate number 4.1.

Table 4.1: Effect of growing media on rate of germination

Growing Medium	Date of sowing	No. of seeds sown	Pot No.	No. of seeds germinate					Total germinated Seeds	%age of germinated seeds
				7th day	14th day	21st day	28th Day	35th day		
Bhal	12.1.11	30	1	2	10	4	4	5	30	100
			2	1	8	1	1	1	14	46.6
			3	-	4	2	1	1	14	46.6
			4	-	5	5	1	1	17	56.6
			5	-	1	3	1	1	9	30
			6	-	1	3	1	1	7	23.3
			7	-	1	-	2	2	6	20
			8	-	3	-	1	1	9	30
			9	-	3	1	1	1	7	23.3
			10	-	2	1	-	-	4	13.3

4.1: Effect of growing media on rate of germination:

Pre-Harvest soil analysis:

Different soil samples were analyzed for their chemical composition by carrying out various tests. Results of chemical analysis of different soil samples before harvesting are shown in the Table 4.2. As evident from the table the pH of the sample decreased with decreasing percentage of bhal and therefore was minimum (pH) in T4 i.e., 50% bhal + 50% organic manure. This showed that addition of organic manure to the bhal shifts the pH towards more acidic but it increased in T1 that was 50% leaf manure. With decreasing pH, electrical conductivity (E.C.) of the soil also decreased in the same manner and increased at last in T1 (50% leaf manure). The amount of carbonates present in m.eq/l of the soil sample also decreases with decreasing amount of bhal in treatments. Similarly the amount of bicarbonates also decreases as shown in table 4.2.

Table 4.2:   Pre Harvest soil analysis

Treatments	pH	EC mS/cm	CO_3^{-2} m.eq/l	HCO_3^{-1} m.eq/l
T0*	8.81	18.66	Nil	12.5
T1*	8.80	3.68	-	5
T2*	8.44	7.5	-	5.25
T3*	8.84	7.6	-	3.3
T4*	9.65	7.23		

T0= 100% bhal

T1=50% bhal + 50% leaf manure

T2= 100% biofert

T3= 50% bhal + 50% biofert

T4= 50% bhal + 50% organic manure.

Effect of different soil composition on shoot length:

The weekly heights of sweet pepper plant were noted for five weeks for five treatments i.e. T0, T1, T2, T3 and T4 as given in the table 4.3 and plate #4.2. The maximum shoot length in T1 i.e. 50% leaf manure + 50% Bhal and minimum in T2 i.e. 100% biofertilizer was recorded.

Table 4.3: Effect of different soil compositions on shoot lengths of sweet pepper

Variety	Date of seedling transplant	Treatment	Shoot length (cm)				
			One week	Two weeks	Three weeks	Four weeks	Five weeks
Sweet pepper	11.3.11	T0*	$4.18^a \pm 1.288$	$4.4^a \pm 1.854$	$6.12^a \pm 3.104$	$7.4^a \pm 5.936$	$7^a \pm 6.079$
		T1*	$3.8^a \pm 0.927$	$3.3^{ab} \pm 1.029$	$5.86^a \pm 1.009$	$6.1^a \pm 1.157$	$6.4^a \pm 8.541$
		T2*	$3.5^a \pm 0.8944$	$3.0^{ab} \pm 1.559$	$2.1^{ab} \pm 2.4$	$1.6^b \pm 2.8$	$0.0^a \pm 0$
		T3*	$3.5^a \pm 1.403$	$9.0^{ab} \pm 1.435$	$1.5^b \pm 43.623$	$0.0^b \pm 0$	$0.0^a \pm 0$
		T4*	$2.7^a \pm 1.596$	$0.8^b \pm 1.8$	$1.2^b \pm 2.4$	$0.0^b \pm 2.61$	$0.0^a \pm 0$
		LSD	1.681	2.750	3.925	4.401	6.818

*T0= 100% bhal, T1=50% bhal + 50% leaf manure, T2= 100% biofert, T3= 50% bhal + 50% biofert, T4=.50% bhal + 50% organic manure

Means followed by different letters in the columns are non significant ranges at P=0.05 according to Duncan's new multiple range tests.

4.2: Effect of different soil compositions on shoot lengths of sweet pepper

Effect of different soil compositions on shoot lengths of sweet pepper

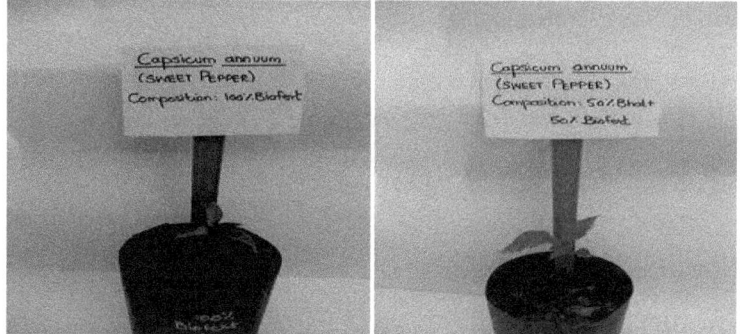

Effect of different soil compositions on number of leaves:

Another parameter for vegetative analysis is the total number of leaves per plant and it was measured weekly along with the shoot length. It was noticed that the number of leaves in different treatments was same initially but it varied with the passage of time. Number of leaves was maximum in T1 (50% leaf manure + 50% Bhal) and was minimum for T2(100% biofert) as shown in the Table 4.4. Plate number 4.3 explains the effect of different soil compositions on number of leaves after two weeks of transplanting.

Table 4.4: Effect of different soil compositions on number of leaves of sweet pepper

Variety	Date of seedling transplant	Treatment	Number of leaves				
			One week	Two weeks	Three Weeks	Four weeks	Five weeks
Sweet pepper	11.3.11	T0*	$5.8^a \pm 1.226$	$5.2^a \pm 1.6$	$8.6^a \pm 3.059$	$8.6^a \pm 4.841$	$7.6^a \pm 3.841$
		T1*	$3.8^a \pm 2.110$	$4.2^{ab} \pm 1.166$	$6.8^a \pm 1.359$	$5.4^a \pm 1.743$	$5.6^a \pm 1.743$
		T2*	$5.4^a \pm 0.8$	$3.0^{abc} \pm 1.6$	$1.4^b \pm 2.4$	$0.0^b \pm 0.0$	$0.0^a \pm 0.0$
		T3*	$5.2^a \pm 0.894$	$1.8^{bc} \pm 1.166$	$1.4^b \pm 0.8$	$0.0^b \pm 0.489$	$0.0^a \pm 0.0$
		T4*	$4.2^a \pm 1.166$	$0.8^c \pm 2.529$	$1.2^b \pm 2$	$0.0^b \pm 0.0$	$0.0^a \pm 0.0$
		LSD	2.166	3.720	3.954	3.476	7.036

*T0= 100% bhal, T1=50% bhal + 50% leaf manure, T2= 100% biofert, T3= 50% bhal + 50% biofert, T4=.50% bhal + 50% organic manure

Means followed by different letters in the columns are non significant ranges at P=0.05 according to Duncan's new multiple range tests.

4.3: Effect of different soil compositions on number of leaves:

Effect of different soil compositions on chlorophyll content

When the crop was eight week old, chlorophyll determination was carried out for sweet pepper plants grown in the five treatments i.e. T0,T1,T2, T3 and T4. Plants were lush green at this stage. Young leaves were chosen for this purpose. The data given in the table 4.5 showed that the amount of total chlorophyll as well as Chlorophyll"a" and Chlorophyll"b" was highest in leaves from all the five treatments. It can also be seen from the table the amount of chl.a and chl.b.

Table 4.5: Effect of different soil compositions on chlorophyll content

Treatments	Chlorophyll content obtained on the basis of fresh weight		
	Chl.a	Chl.b	Total chlorophyll
T0*	$120.4^b \pm 0.329$	$210.3^b \pm 0.418$	330.7
T1*	$125.4^a \pm 0.294$	$222.3^a \pm 0.169$	347.7
T2*	-	-	-
T3*	$120.3^c \pm 0.368$	$210.3^c \pm 0.205$	330.6
T4*	$120.3^c \pm 0.249$	$135.5^c \pm 0.286$	255.8
LSD	0.288	0.185	

*T0= 100% bhal, T1=50% bhal + 50% leaf manure, T2= 100% biofert, T3= 50% bhal + 50% biofert, T4=.50% bhal + 50% organic manure

Means followed by different letters in the columns are non significant ranges at P=0.05 according to Duncan's new multiple range tests.

Effect of different fertilizer composition on shoot length:

The weekly heights of sweet peppers plant were noted for five weeks for nine treatments i.e. Phostrogen (2.75, 3.75, 4.75), NPK (0.927, 1.85,3.75), Hoagland solution (50,100,150). one pot was kept as control. The data given in the table 4.6 and plate number 4.4 showed maximum shoot length i.e. $40.4^a \pm 0.374$ in NKP 3.75 which may be due to the fact that NPK provided the plants with ideal concentrations of Nitrogen, Phosphorus and Potassium relative to Phostrogen and Hoagland solution that gave the minimum shoot length i.e. $24.2^c \pm 0.216$ & $17.433^c \pm 0.1699$, respectively.

Table 4.6: Effect of different fertilizer composition on shoot length

Variety	Treatment	Shoot length (cm)		
		One week	Two weeks	Three weeks
Sweet pepper	P 2.75	$13.266^c \pm 0.205$	$14.266^c \pm 0.355$	$14.333^c \pm 0.374$
	P 3.75	$18.5^a \pm 0.244$	$28.233^a \pm 0.420$	$27.266^a \pm 0.205$
	P 4.75	$17.5^b \pm 0.286$	$26.666^b \pm 0.205$	$24.2^b \pm 0.216$
	LSD	0.947	0.937	0.632
	NPK 0.927	$17.266^c \pm 0.309$	$19.333^c \pm 0.385$	$24.4^c \pm 0.402$
	NPK 1.85	$14.466^b \pm 0.205$	$17.3^b \pm 0.094$	$20.466^b \pm 0.368$
	NPK 3.75	$26.4^a \pm 0.368$	$36.26^a \pm 0.309$	$40.4^a \pm 0.374$
	LSD	0.413	0.185	0.261
	H 50	$18.366^a \pm 0.385$	$19.533^a \pm 0.294$	$26.466^a \pm 0.3299$
	H 100	$16.366^b \pm 0.386$	$17.3^b \pm 0.339$	$20.366^b \pm 0.368$
	H 150	$15.366^c \pm 0.329$	$16.3^c \pm 0.2449$	$17.433^c \pm 0.1699$
	LSD	0.858	0.528	0.119

NPK=Nitrogen. Phosphorus, Potassium

P= Phostrogen

H= Hoagland solution

Means followed by different letters in the columns are non significant ranges at P=0.05 according to Duncan's new multiple range tests.

44

4.4: Effect of different compositions NPK fertilizer on shoot length:

Effect of different compositions of Phostrogen on shoot length

 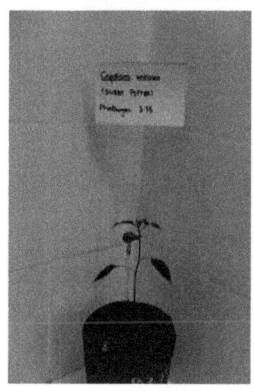 

Effect of different compositions of Hoagland solution on shoot length

Effect of different fertilizer compositions on number of leaves

Another parameter for vegetative analysis is the total number of leaves per plant and it was measured weekly along with the shoot length. It was noticed that the number of leaves in different treatments was same initially but it varied with the passage of time. The number of leaves was recorded maximum i.e. $38.2^a \pm 0.216$ in NPK 3.75. As NPK provided ideal concentration of nitrogen for the maximum growth of leaves. The minimum number of leaves were recorded i.e. $26.3^b \pm 0.294$ in H 150 and $24.4^c \pm 0.326$ in P 2.75, respectively. Plate number 4.5 explains the effect of different soil compositions on number of leaves after two weeks of transplanting.

Table 4.7: Effect of different fertilizer compositions on number of leaves

Variety	Treatment	Number of leaves		
		One week	Two weeks	Three Weeks
Sweet pepper	P 2.75	$13.5^a \pm 0.3681$	$27.4^a \pm 0.329$	$30.4^a \pm 0.205$
	P 3.75	$11.3^b \pm 0.287$	$22.5^b \pm 0.294$	$28.3^b \pm 0.368$
	P 4.75	$9.3^c \pm 0.249$	$16.3^c \pm 0.368$	$24.4^c \pm 0.326$
	LSD	0.560	0.206	0.244
	NPK 0.927	$20.4^c \pm 0.286$	$22.3^c \pm 0.294$	$16.4^c \pm 0.374$
	NPK 1.85	$24.3^b \pm 0.205$	$24.4^b \pm 0.124$	$24.4^b \pm 0.408$
	NPK 3.75	$25.2^a \pm 0.287$	$37.3^a \pm 0.249$	$38.2^a \pm 0.216$
	LSD	0.658	0.450	0.459
	H 50	$22.3^a \pm 0.294$	$27.4^a \pm 0.286$	$33.2^a \pm 0.204$
	H 100	$21.4^b \pm 0.294$	$24.4^b \pm 0.169$	$26.4^b \pm 0.249$
	H 150	$18.3^c \pm 0.245$	$24.2^b \pm 0.326$	$26.3^b \pm 0.294$
	LSD	0.206	0.244	0.177

NPK=Nitrogen. Phosphorus, Potassium

P= Phostrogen

H= Hoagland solution

Means followed by different letters in the columns are non significant ranges at P=0.05 according to Duncan's new multiple range tests.

4.5: Effect of different fertilizer compositions on number of leaves:

Effect of different fertilizer compositions on chlorophyll content:

When the crop was eight week old, chlorophyll determination was carried out for sweet pepper plants grown in the nine treatments. Plants were lush green at this stage. Young leaves were chosen for this purpose. The data given in the table 4.5 showed that the amount of total chlorophyll as well as Chlorophyll"a" and Chlorophyll"b" was highest in leaves from the pot treated with NPK 3.75 i.e. 336.2, 152.7^a±0.368 and 184.5^a±0.249, respectively.

Table 4.8: Effect of different fertilizer compositions on chlorophyll content

Treatments	Chlorophyll content obtained on the basis of fresh weight		
	Chl.a	Chl.b	Total chlorophyll
P2.75	$62.2^a\pm0.374$	$72.7^a\pm0.386$	134.9
P 3.75	$49.7^a\pm0.249$	$53.2^b\pm0.368$	102.9
P 4.75	$48.9^a\pm0.205$	$51.0^b\pm0.339$	99.9
LSD	19.107	18.527	-
NPK 0.927	$151.1^a\pm0.339$	$184.9^a\pm0.374$	237
NPK 1.8	$151.9^a\pm0.339$	$184.6^a\pm0.228$	336.5
NPK 3.75	$152.7^a\pm0.368$	$184.5^a\pm0.249$	336.2
LSD	0.684	0.761	-
H 50	$130.5^a\pm0.244$	$141.6^a\pm0.249$	272.1
H 100	$119.1^a\pm0.329$	$119.4^a\pm0.205$	238.5
H 150	$48.3^b\pm0.329$	$43.4^a\pm1.22$	91.7
LSD	25.464	101.681	-
Control	119.464	17.391	136.85

NPK=Nitrogen. Phosphorus, Potassium

P= Phostrogen

H= Hoagland solution

Means followed by different letters in the columns are non significant ranges at P=0.05 according to Duncan's new multiple range tests.

Post harvest analyses of plants:

After chlorophyll determination, final harvest analysis of the crop or plants was carried out. It was carried out after twelve weeks. Final harvest analysis was carried out to determine the effect of different fertilizer combinations on their growth. Plants from N.P.K were healthier and better developed than from Hoagland and Phostrogen. Shoot length, root length, fresh weight of shoot and root and dry weight of shoot and root and dry weight of shoot and root were maximum in NPK 3.75.

Table 4.9: Post harvest analyses of plants

Treatment	Root length (cm)	Shoot length (cm)	Fresh weight root (g)	Fresh weight shoot (g)	Total fresh weight (g)	Dry weight root (g)	Dry weight shoot (g)	Total dry weight (g)
P 2.75	5.566^a±0.140	14.433^a±0.205	8.523^a±0.26	14.883^a±0.522	18.45	1.058^a±0.08	7.893^a±0.157	4.34
P 3.75	5.563^a±0.085	12.766^a±0.294	8.504^c±0.14	5.263^a±0.089	11.93	1.12^a±0.054	7.753^a±0.154	1.89
P 4.75	3.05^a±0.08	14.0^a±0.329	0.729^a±0.135	4.133^a±0.08	22.29	1.89^a±4.6	0.74^a±0.01	2.34
LSD	5.121	16.649	176.968	12.976^a		14.823	3.264	
NPK 0.927	5.56^c±0.054	12.8^c±0.33	3.6^a±0.26	32.28^b±0.163	13.27	0.642^a±0.089	8.86^a±1.2	4.33
NPK 1.85	5.56^c±0.054	28.7^c±0.345	4.5^a±0.2	27.34^a±1.03	35.08	1.8^a±0.109	8.56^a±0.054	3.06
NPK 3.75	6.1^a±0.089	35.4^a±0.219	8.32^a±0.22	36.2^a±1.25	15.38	2.56^a±1.23	11.07^a±0.283	2.56
LSD	0.355	0.874	0.684	0.616		0.355	0.526	
H 50	8.86^b±0.17	19.4^a±0.219	6.4^a±0.219	31.4^b±0.167	15.06	1.14^b±0.08	4.14^a±0.08	2.06
H 100	22.2^c±0.140	12.0^b±0.245	2.34^b±0.109	25.6^a±0.260	27.1	1.5^a±1.2	3.2^a±0.16	4.22
H 150	15.36^a±0.522	6.3^c±0.179	4.24^b±0.109	22.3^a±0.268	15.45	0.82^b±0.109	2.2^a±0.109	4.15
LSD	1.080	0.645	0.576	0.814		0.313	0.414	
Control	15	30	3.195	22.792	25.987	0.866	3.478	4.344

Means followed by different letters in the columns are non significant ranges at P=0.05 according to Duncan's new multiple range tests.

4.6: Post harvest analysis of plants

Post harvest analysis of plants

Post harvest analyses of plants

Post harvest soil analyses:

Post harvest soil analysis of the nine compositions was carried out for the same tests as were done for pre-harvest soil analysis. Results have been shown in the table 4.8. As shown in the table, pH and E.C.were minimum in NPK i.e. NPK 3.75 and maximum in Hoagland solution.

Table 4.10: Post harvest soil analyses

Variety	Treatment	pH	EC mS/cm	CO_3^{-2} m.eq/l	HCO_3^{-1} m.eq/l
	Control set	9.05	12.7	-	5
	P 2.75	9.12	12.2	-	3
	P 3.75	9.20	11.7	6	2
	P 4.75	9.29	11.3	4	2.5
Sweet pepper	NPK 0.927	8.99	1.4	-	1.4
	NPK 1.85	8.98	2.3	-	2.1
	NPK 3.75	8.95	1.6	-	5.8
	H 50	9.66	6.5	-	5.6
	H 100	9.64	7.8	11	5.5
	H 150	9.62	7.88	-	4.0

NPK=Nitrogen. Phosphorus, Potassium

P= Phostrogen

H= Hoagland solution

Means followed by different letters in the columns are non significant ranges at P=0.05 according to Duncan's new multiple range tests.

Discussion

The present study deals with the various aspects of different levels of soil combinations on the vegetative and reproductive growth of *Capsicum annuum* L. Firstly percentage germination was observed in growing media i.e. bhal. Similarly pre and post harvest soil analysis was done with various parameters i.e. pH, electrical conductivity, estimation of carbonates and bicarbonates was calculated. Various other parameters such as number of leaves and plant height per plant were noticed every week. Chlorophyll determination and the post harvest analysis of the plants were also done.

Our results in the table 4.2 and plate number 4.1 showed the pre harvest soil analysis of sweet pepper plants. As evident from the Table 4.2 that the pH of the sample decreased with decreasing percentage of Bhal and therefore was minimum pH was recorded in T4 (50%bhal + 50% organic manure). The decrease in pH is may be due to the fact that addition of organic manure to the Bhal shifted the pH towards more acidic. It is evident from the study that there exists a direct correlation among pH and electrical conductivity (Amor, 2007) as from data, with decreasing pH, the electrical conductivity of soil also decreased. Carbonates and bicarbonates in meq/l of soil sample also decreased with decreasing amount of Bhal. Amor, (2005) reported that the effect of organic manure (ecological organic fertilizer) different complex substrate on hot pepper growth. The complex substrate of ecological organic manure, vermiculite and turf (1:1:1), or ecological organic fertilizer and vermiculite (1:1) improve growth of hot pepper. It mainly increasing plant height, stem diameter, leaf numbers and photosynthesis, making flowering and maturing earlier biologic and economies yield higher.

Weekly growth assessment of sweet pepper covered various growth parameters in the present work such as plant height, number of leaves, total chlorophyll content including chlorophyll a & chlorophyll b, root length, shoot length, fresh and dry weight etc. in the tables 4.3, 4.4 & 4.5. All the growth parameters showed maximum values for T1(50%bhal + 50% leaf manure).The study showed that the plants transplanted in leaf manure showed better growth with highest number of leaves and increased plant growth. Serrano *et al.,* (2009) reported that when sweet pepper plants were grown in a greenhouse under three different cultivation methods (organic, integrated and conventional farming). During the crop cycle, plant growth and especially yield and fruit quality parameters were monitored to determine the effects of the different fertilization strategies. Plant fresh weight and total leaf fresh weight were progressively reduced, relative to the other treatments, in the organic treatment compared with the conventional, and at the end of the crop cycle these parameters were reduced by 32.6 and 35% respectively. Fruit firmness, pH and total soluble solids content showed higher values with the organic method, but these differences were not significant with respect to the conventional method.

It has been shown in the table 4.5 that extraction of chlorophyll from the fresh leaves of sweet pepper grown in T1 is highest followed by plants grown in T0, T3 and T4. An increase in chlorophyll content with T1 treatment was also reported by Marin, 2008. Plant growth parameters such as shoot dry matter, total leaf area and leaf weight fraction were all reduced in the organic treatment compared with the conventional. Leaf expansion was dramatically reduced in the organic treatment 155 days after transplanting. Relative growth rates were significantly affected by nitrogen concentration in each organ and were directly to the cultivation method. Chlorophyll $(a + b)$ contents in the leaves were reduced in the organic treatment and were directly correlated with the nondestructive quantification of chlorophyll using a portable chlorophyll meter. Net photosynthesis was also reduces in the organic treatment, but chlorophyll fluorescence was not affected. This study shows that biometric monitoring and first & non-invasive techniques of plant nutrient status analyses could help to improve growth as easy and useful tools to follow the nutrient status at different phonological stages, especially when no chemical fertilizers are applied.

The present study reveals that growth of sweet pepper plant is maximum in pots which are fertilized with the NPK 3.75 (table 4.6) relative to Phostrogen and Hoagland solution. This is may be due to fact that soluble NPK provided the ideal concentration of nitrogen, phosphorus and potassium in soluble forms to the plants. This easily contributed for the maximum growth and yield of the plants of sweet pepper. The same results were reported by Amor *et al.,* 2010; Gopinath *et al.*, 2009.

Weekly growth assessment covered various growth parameters in the present work such as plant height, number of leaves, root, shoot fresh and dry weight etc. in the tables 4.7, 4.8, 4.9 & 4.10. All the growth parameters showed maximum values for NPK 3.75 (Awodun *et al.,* 2007; and Sui *et al,* (2002). Growth and yield parameters such as number of leaves and branches, plant height, stem girth and number were significantly (p>0.05) increased by NPK treatment. However NPK fertilizer increased soil N, P and K status leaf N status and growth and yield parameters compared with manure treatments.

Salama and Zake, (2000) also studied the effect of fertilization on the qualities of sweet pepper (*Capsicum annuum* L.) cultured in greenhouse, the experiment was carried out by drop irrigation. High fertilization of N, P and K had no negative influence on content of nitrate and total soluble sugars in fruit. Decreasing fertilization would cause increasing of the inosital and sucrose contents. Glucose and fructose were the main soluble sugars in fruit. The metabolism of carbohydrates was weakened by high fertilization, and it expressed as restraining the accumulation of starch content in leaf and perhaps, it was due to the higher content of chlorophyll in the leaf.

REFERENCES

Alabi, D. A. 2005. Effects of fertilizer phosphorus and poultry droppings treatments on growth and nutrient components of pepper (*Capsicum annuum* L). *African Journal of Biotechnology.* **5** (8). 671-677.

Amor, F. M. D. 2007. Yield and fruit quality response of sweet pepper to organic and mineral fertilization. *Renewable Agriculture and Food Systems.* **22** (3). 233-238.

Amor, F. M. D., Paula, C. C., David, J. W., Jose, M. C and Ramon, M. 2010. Effect of foliar application of antitranspirant on photosynthesis and water relations of pepper plants under different levels of CO_2 and water stress. *Journal of Plant Physiology.* **167** (15). 1232-1238.

Amor, F. M. D., Serrano, M. A., Fortea, M. I., Legua, P. and Nunez, D. E. 2008. The effect of plant-associative bacteria (*Azospirillum* and *Pantoea*) on the fruit quality of sweet pepper under limited nitrogen supply. *Scientia Horticulturae.* **117** (3). 191-196.

Amor, F.M.D. 2005. Growth, photosynthesis and chlorophyll flourescenceof sweet pepper plants as affected by the cultivation method. *Annals of applied biology.* **148** (2). 133-139.

Arnon, D., 1949. Copper enzymes in isolated chloroplasts. Polyphenol oxidase in *Beta vulgaris. Plant physiol.* **24**. 1-15.

Awodun, M. A., Omonijo, L. I. and Ojeniyi, S.O. 2007. Effect of Goat Dung and NPK Fertilizer on Soil and Leaf Nutrient Content, Growth and Yield of Pepper. *International Journal of Soil Science.* 2 (2). 142-147.

Berova, M. and Karanatsidis, G. 2009. Influence of Bio-fertilizer produced by *Lumbricus rubellus* on growth, leaf gas-exchange and photosynthetic pigment of pepper plants (*Capsicum annuum* L.). *Acta Hort.* (ISHS). **830**. 447-452.

Boukhenfouf, W. and Ahmed, B. 2011. The radioactivity measurements in soils and fertilizers using gamma spectrometry technique. *Journal of Environmental Radioactivity.* **102** (4). 336-339.

Burt J. 2005. Growing capsicums and chillies. Farmnote. Department of Agriculture and Food.

61

Dileep, S. N. and Sasikala, S. 2009. Studies on the effect of different organic and inorganic fertilizers on growth, fruit characters, yield and quality of chilli (*Capsicum annuum* L.) cv. K-1. *International Journal of Agricultural Sciences*. **5** (1). 229-232.

El-Al. F.S.A., 2009. Effect of urea and some organic acids on plant growth, fruit yield and its quality of sweet pepper (*Capsicum annum*). *Research Journal of Agriculture and Biological Sciences*. **5** (4). 372-379.

Gopinath, K . A., Supradip, S., Mina, B. L., Harit, P., Srivastva, A. K. and Gupta, H. S. 2009. Bell pepper yield and soil properties during conversion from conventional to organic production in Indian Himalayas. *Scientia Horticulturae*. **122** (3). 339-345.

Gowda, K. K., Mahantesh, S. and Sreeramu, B. S. 2002. Effect of bio-fertilizers with graded levels of nitrogen and phosphorus on growth, yield and quality of chillies (*Capsicum annuum* L.) cv. Byadagi Dabba. *Book chapter, Conference paper Proceedings of the 15th Plantation Crops Symposium Placrosym*. 304-309.

Irfan, Y.S. and Khalid, A. N. 2007. In vitro biological control of Fusarium oysporum-r causing wilt in capsicum annuum. Mycopath. **5** (2). 13-30.

Kotur, S. C., Ramesh, P. R., *Anjaneyulu, K. and Ramachandran, V.* 2010. Direct and residual effect of nitrogen in high-yielding/hybrid chilli (*Capsicum annuum*) radish (*Raphanus sativus*) system. *The Indian Journal of Agricultural Sciences*. **80** (7).

Lindley, J. 1838. About *Capsicum annuum*. *Flora Medica* .509.

Marcussi, F. F. N., Roberto, L. V. B., Leandro, J. G. G. and Rumy, G. 2004. Macronutrient accumulation and partioning in fertigated sweet pepper plants. *Scientia Agricola*. **61** (1).

Marin, A., Maria, I. G., Pilar, F., Pilar, H. and Maria, V. S. 2008. Microbial Quality and Bioactive Constituents of Sweet Peppers from Sustainable Production Systems. *Journal of agriculture and food chemistry*. **56** (23). 11334–11341.

Mc Keague, J.A 1978. Mannualonsoil sampling and methods of analysis. *Canadian society of soil science*. 66-88.

Mc Lean, E.O. 1982. Soil pH and lime requirement. Methods of soil analysis part 2: Chemical and microbial properties. *Am. Soc. Argon. Madison. WI. USA*. 199-224.

Melo, A.M.T. 1997. Analise genetica de carateres de fruto em hibrios de pimentao. *ESALQ/USP* 112.

Palada, M.C. and Wu, D.L. 2006. Evaluation of Chili rootstocks for grafted sweet pepper production during tha hot-wet and dry-hot season in Taiwan. *Acta Hort. (ISHS)*. **767**. 151-158.

Pascual, I., Azcona, I., Aguirreolea, J., Morales, F, Corpas, F.J., Palma, J.M., Rellán-Alvarez, R. and Sánchez-Díaz, M. 2010. Growth, yield, and fruit quality of pepper plants amended with two sanitized sewage sludges. *Journal of agriculture and food chemistry*. **58** (11). 6951-9.

Russo, V. M. 2005. Organic vegetable transplant production. *Hort. Science*. **40** (3). 623-628.

Russo, V.M. 2010. Irrigation timing and fertilizer rate in peppers. *Proceedings of Horticultural Industry Show*. 157-158.

Richards, L. A. 1954. Diagnosis and improvement of saline and alkaline soil. *USDA. Agric Handbook* 60. Washington D.C.

Salama, G. M. and Zake, M. H. 2000. Fertilization with manures and their influence on sweet pepper of plastic-houses. *Annals of Agricultural Science*. **38** (2). 1075-1085.

Serrano, M., Pedro, J. Z., Salvador, C., Fabián, G., Domingo, M. N. and Daniel, V. 2009. *Food Chemistry*. 118 (3). 497-503.

Serrano, R. A., and Guerra-Sanz, J. M. 2006. Quality fruit improvement in sweet pepper culture by bumblebee pollination. *Scientia Horticulturae*. **110**. 160-166.

Sui, F., Wang, Y., Nagatomo, M., Chishaki,d N., Wunimuren and Inanagea, S., 2002. Effect of fertilization on the qualities of sweet pepper in greenhouse culture. *Journal of applied ecology*. **13** (1). 63-66.

- www.avrdc.org/LC/pepper/swtprod/03int.html
- www.avrdc.org/LC/pepper/swtprod/09fert.html
- www.britannica.com
- www.calantilles.com
- www.calantilles.com/capsicum_pepper
- www.da.gov.ph/tips/sweetpepper.html
- www.debco.com.au/products/fertilisers/phost.php
- www.myfolia.com/capsicum
- www.nal.usda.gov/fnic/foodcomp/search
- www.thechileman.org/guide-fertilizer.php
- www.whfoods.com/bellpepper
- www.wisegeek.com
- www.wisegeek.com/capsicum
- http://aggie-horticulture.tamu.edu/archives/parsons
- http://en.wikipedia.org/wiki/Hoagland_solution
- en.wikipedia.org/wiki/capsicum
- en.wikipedia.org/wiki/capsicum_annuum
- www.answer.com/topic/chilli-pepper.